Top is Ant's pal.

"See me dip and zip," said Top.

"I like to go fast," he said.

"No. Stop it!" said Ant.
"I like to dip and zip.
But a top spins and slips."

Top was so mad.

He said, “I can dip and zip.”

Ant said, “No, you cannot!”

Ant is as mad as Top.

The two of them yell.

Then Ant and Top split up.

Top sits in his box.

Top is still mad.

Ant lands on a plant.

Ant is still mad.

Ant is a good pal.
He plans to make up.
"Stop," said Ant at last.
"You can dip and zip."

Top zips and dips.

"I did a zip!" said Top.

"It is a snap, and so is a dip!"

Ant said, "Go, Top, go!"

The two pals zip and dip.
Then they spin and slip.
They have so much fun!